Chirality and Life

Dr. Sunil T. Patel

Dr. Asha D. Patel

Dr. Naresh K. Prajapati

Canadian Academic Publishing
2013

Price : $27.86

First Edition : 2013

ISBN : 978-0-9921651-3-0

Publisher ISBN Prefix : 978-0-9921651

ISBN Allotment Agency : Library and Archives Canada (Govt. of Canada)

Published & Printed by
Canadian Academic Publishing
81, Woodlot Crescent,
Etobicoke,
Toronto, Ontario, Canada.
Postal Code- M9W 6T3
Phone- +1 (647) 633 9712
http://www.canadapublish.com

PREFACE

Chirality describes the chemical, pharmacological, biological, Physical, mathematical science, observations, events & experiences. We see these all in our life by a lot in different manner. We can not express the chirality that "how much important on the paper" to investigate hither-to unexplained and undiscovered phenomena still.

This book inspire to investigator & to present the new development for world.

The writing of this book is inspired from one of turing corner of my life. I try to include almost all the field related to asymmetry. But this is only primary information about Chiral world.

This book has both the background academic as well as industrial. The philosophy of the book create views in mind of researchers. So the applicative thoughts in industrial life remain unceased to human especially in pharmaceuticals, food, flavors & fragrances.

The information divided into six chapters. Chapter-1 include the history & which event put the word 'Chirality'. Chapter-2 devotes the definition of chirality after the proved things. Also where it is applied & our attention for daily life in molecular world. Chapter-3 explains the enantiomers for different molecule.where amino acids & carbohydrates like natural compounds. Chapter-4 especially explain for homochirality in molecular biology with of receptor based cell in human body.Chapter-5 content the chirality in industrial aspect with it's need. Also how to get by some methods.Chapter-6 shows the importance in real world of chirality of different science by scientists throughout the various laboratories & places.

Acknowledgment

First of all, I would like to thank my friend Dr. K.M.Joshi, A principal of Govt. college for his constructive criticism & continuous comments to inspire for book writing for systematic development of my academic career.

I like to acknowledge my co-authors Dr. A. D. Patel & Dr.N.K.Prajapati for their help adding his expertise to shape this book & financial assistance to publish this book.

I wish to give a fame who allow to spare the time along with my social responsibilities and a profession of mine. And those are my family members. I thank of them a major.

I am also thankful to a chief of editor, Mr. Chintan Mahida giving a systematic information to publish the book in very easy way with his art of work & business.

- *Dr. Sunil T. Patel, Dr. Asha D. Patel, Dr. Naresh K. Prajapati*

CONTENTS

CHAPTER – I
HISTORY OF CHIRALITY

1.1 How event of Chirality identified :

The word 'Chirality' has fascinated scientists for over 150 years. Chirality is hardly a new concept in chemistry, yet advance in daily life. Chirality in nature came to light only some one hundred and fifty years ago.

By the way...... please pronounce "ky-rality" with a "K", like chiropractor. The Greek name kheir means "hand." So, chirality indeed means "handedness." Following Louis Pasteur's discoveries, the name was coined by the Irish physicist William Thomson, alias Lord Kelvin.

It was Louis Pasteur, the great French scientist who discovered chirality in the spin of molecules in 1848. It is a great tribute to his exceptional intelligence and keen observation that while examining a certain salt of tartaric acid he noticed that there were two types of crystals, each a mirror image of the other. He carefully separated the two, dissolved them in water and made a beam of light pass through the solution. He was surprised to discover that the polarized light was rotated differently by the two specimens. One was rotated clockwise and the other anticlockwise. This clearly meant that the molecules of the two separated specimens of tartaric acid were either spinning to the right or to the left-neither could be superimposed on the other through the beam of plane polarized light only. This was the first ever case of chirality observed by scientists at the elemental level.

Another singularly significant discovery in the same field was made yet again by Pasteur in 1857. One day he noticed the growth of a mould in a chemical solution lying in a jar. Instead of throwing away the solution as contaminated, he made a beam of light pass through it to examine the effect, if any of that mould on the solution. He was astounded to discover that the solution though inactive in relation to light prior to its contamination had suddenly become active and started polarizing light. It was inactive in relation to light for the simple reason that it was composed of an equal number of right spinning and left spinning molecules each neutralizing the other's effect on light. Hence the polarity displayed by the contaminated specimen could only mean that the mould had eaten up only such molecules as spun in the same direction and left completely untouched those which spun in the opposite. One mystery was thus resolved but only after having given birth to another much more complex one. How could a mere mould detect the spin of molecules with such unfailing exactness and why was it at all partial to the molecules spinning in any specific direction? These were the questions which baffled the mind of Pasteur then and still baffle the minds of scientists today. For how long they remain unanswered, the scientists know not. The magnitude of the dilemma is enormous. The molecules of any element or compound, right spinning or left spinning share exactly the same chemical and physical properties. What or who dictates their propensity to spin in any particular direction is a brain-twister enough but when it comes to the most uncanny ability of life to detect which molecules are spinning in which direction, the question acquires bizarre astronomical proportions. None of the five senses bestowed to man are equipped with any known mechanism which can determine the spin of molecules. The spinning molecules leave no imprint on the property of matter to become detectable through human sensory organs. But what of moulds which have no known sensory organs; all they have is a diffused sense of awareness?

The chirality of the observable world provides inspirable and impetus for many domains of human endeavor. The 'handedness' of both the macroscopic and molecular worlds is exorably tied to our experience and our quest to understand the nature.

Today the echo of pasture's landmark revelation of the molecular origins of optical activity still resonates in the research activities of many scientists. But instead of rationalizing observable phenomena, Chemists are now engaged in the design and synthesis of molecular substances that will display specific properties, activities and functions. In so far as all of these activities require access to chiral molecules in stereochemically defined form and high configurational purity. It is no wonder that asymmetric synthesis has taken on a central role as the enabling discipline that can provide the materials and methods for the manifold applications of chiral compounds.

Vladimir Prelog says that *"The World is Chiral and clinical, enjoy symmetry wherever you find it"*. Some common examples are given by followings:

The number and variety of chiral objects and situations that one can find in daily life. Gloves & Shoes are designed to fit only one hand or foot. Golf clubs and many musical instruments are designed to be right or left handed. Doors swing open to the right or the left. Screws, bolt & nuts are threaded in right or left handed manner. Automobiles are also chiral (steering wheel is on left hand side in U.S. and right hand side in Great Britain).Cloths also (Men's shirt button left over right and women`s shirts button right over left).

Chiral (Asymmetry) is possible in many molecule possess molecular symmetry incompatible with crystal symmetry. e.g. Ferrocene. Ferrocene molecule possess 5-fold rotational symmetry but crystal can't impose it on the molecule (A crystal is a 3-dimentional structure generated by translational of simple units known as unit cell along each of 3 principle axes of the crystal)

This amazing tale of chirality in nature does not end here. It just begins. Since the time of Pasteur, research on chirality has made tremendous progress and many more extremely perplexing examples have come to light testifying that chirality can be unmistakably detected by different species of life.

By now chirality is discovered to operate at every level of material existence. Yet the manner of how and why it so behaves is far from understood. Until 1957 it was believed that the four fundamental forces which govern the interaction of elementary particles were parity conserving. This simply means that all particles at elementary level had chiral-symmetry. However, in 1957 Chien-Shiung Wu and her colleagues at Columbia University discovered that beta particles emitted from radioactive nuclei did not display chiral-symmetry. The left-handed electrons far out numbered the right-handed ones. It was further discovered that the tiniest subatomic particles, neutrinos and anti-neutrinos which are electrically neutral and move at the speed of light also display a certain spin. But unlike electrons which predominantly prefer left-handed spin, anti-neutrinos are always partial to the right-hand. The contrary is not found in nature. No one knows

4

why chiral-asymmetry exists at such fundamental levels of existence at all.

Many hypotheses are being presented but most are found to be simply preposterous when examined more minutely. However, there is one suggestion which seems to have provided scientists with a clue to the factor possibly at work at the most rudimentary level of chirality in nature. Yet at this level, it is too ethereal to be demonstrated or verified. It is related to a theory which unifies the weak and electromagnetic forces first propounded by Dr. Abdus Salam, Steven Weinberg and Sheldon Glashow in 1960. That theory predicted a new electroweak force which does not conserve parity. This disparity according to scientists could possibly be responsible for the right-handed spin of antineutrinos and left-handed spin of neutrinos as well as that of electrons. But this weak electric force cannot be contemplated as the causative factor to produce the right sided or left sided behaviours at all other levels of chirality. The behavioural difference between the two sometimes perplexes scientists, particularly in relation to the role they play in biotic evolution. The problem is further compounded when we observe that the two right sided and left sided components of exactly the same chemical formula exert a completely different influence on life in odd ways. The following are some fascinating examples:

Limonene is a compound found both in lemons and oranges. There is not the slightest difference in their chemical formula, yet the spin of limonene molecules in lemons is invariably opposite to the molecular spin of limonene found in oranges. Limonene in lemons is always right spinning while in oranges it is always left spinning. How on earth could lemons and oranges always pick the limonene of a specific spin for their consumption while the difference between their limonene is merely that of molecular spin ? It needs to be emphasized yet again that both the right sided and left sided specimens of limonene contain exactly the same chemical and physical properties. How the olfactory glands of the human nose can ever detect the difference of the spin in oranges and lemons and ascribe to them completely different smells is absolutely astounding. Of course there has to be some reason but as yet we cannot identify it.

Another example relates to the influence of chirality on life of a rather sinister nature. This came to light in 1963 when a drug, thalidomide was introduced by a pharmaceutical company for the cure of morning sickness in pregnant women. Many were cured but for many others it proved disastrous. Horrible congenital defects were found in the babies born to some mothers treated by the same drug. A subsequent intensive research revealed that the pharmaceutical company which manufactured thalidomide had inadvertently manufactured two types of thalidomide compounds of the same formula. While the molecules of one type spun in one direction, those of the second type spun in the other. While one type cured morning sickness without producing any adverse effect on the embryo, the other type produced the most horrible congenital deformities instead of curing the morning sickness. The most profound side-effect was the deformities of the lower limbs among the infants born under its influence.

Another intriguing case of the detection of the spin and the preference of one spin over the other is found at the most fundamental level of life. Although there were several hundred amino acids freely available in the primordial soup from which such proteins were created as made the fundamental bricks of life (DNA and RNA), 'nature' selected only twenty amino acids out of them and they were all left spinning!

In the case of selecting molecules for building sugars however, the choice was reversed. The molecules of all the four different forms of sugars responsible for the provision of energy to all forms of life are, without exception, right spinning. This means that all natural sources of sugar available to life like sugar cane, beet root, fruit, etc., manufacture sugar consisting only of right spinning molecules.

Nevertheless, a successful experiment was conducted a few years ago for synthesizing sugar comprising only left spinning molecules. It was discovered that this artificially synthesized sugar, though exactly the same in taste, chemical properties and cooking behavior was totally rejected by the human digestive system. Not a molecule was assimilated. This gave rise to the bizarre idea of manufacturing sugar consisting of only left spinning molecules on a commercial scale not only for the benefit of diabetics but also for the pleasure of gourmands and gluttons. They could consume mountains of sugar without the fear of accumulating even a molehill of fat. The only snag is that at present the cost of manufacturing left spinning synthetic sugar is prohibitive. A mountain of money would be needed to produce a mere molehill of such sugar. Perhaps only the royal highnesses of oil rich monarchies sitting upon mountains of oil wealth could afford this luxury.

The apparently arbitrary preference for right or left also manifests itself in many other ways. Most humans are right-handed and the arrangement of the heart and liver is universally left sided and right sided respectively, barring a few rare individual exceptions of course.

Roger A. Hegstrom and Dillip K. Kondepudi in their jointly authored article "The Handedness of the Universe" published in Scientific American, January 1990, present many examples of handedness in nature without any apparent reason for preference. While observing that most people are right-handed, they fail to recognize any reason

'...why right- and left-handed persons are not born in equal numbers.'

But it is not a prerogative of the human race alone to display definite trends with regards to handedness.

On partiality to sidedness as found in the animal kingdom and vegetative behaviour, they write:

'Right-handed or dextral shells dominate-on both sides of the Equator. Among these right-dominated animals, left-handed individuals exist only as a result of mutations, which appear with a frequency ranging from about one in hundreds to one in millions, depending on the species.'

right sided grooves   **left sided grooves**

In contrast to them, the lightning-whelk of the Atlantic coast are

predominantly left-handed. In plants, the honeysuckle winds around its support in a left-handed helix while the bindweed prefers winding from right to left. Even in bacteria some of their colonies spiral from right to left yet as the temperature increases they reverse the spiral direction to left-handed turns.

These are but a few cases. At every level of evolution we find many other outstanding examples of how life displays partiality to the spin of molecules. Their study excites wonderment and leaves one bewildered. There has to be a Conscious All-Wise Supreme Selector who made choices at every stage of decision making or one has to ascribe this role to the haphazard vagaries of blind nature !

1.2   An Abbreviated Ancient History

The term optical activity is derived from the interaction of chiral materials with polarized light. A solution of the (−)-form of an optical isomer rotates the plane of polarization of a beam of polarized light in a counter clockwise direction (levorotatory), vice-versa for the (+) (dextrorotatory) optical isomer. The property was first observed by Jean-Baptiste Biot in 1815 and gained considerable importance in the sugar industry, analytical chemistry and pharmaceuticals. Louis Pasteur deduced in 1848 that this phenomenon has a molecular basis. Artificial composite materials displaying the analog of optical activity but in the microwave region were introduced by J.C. Bose in 1898 and gained considerable attention from the mid-1980's. The term chirality itself was coined by Lord Kelvin in 1873.

1.3 A memorial historical events:

Inception (1815 - 1905)

In 1815 the French Physicist Jean-Baptiste Biot showed that certain chemicals could rotate the plane of a beam of polarised light, a property call optical activity. The nature of this property remained a mystery until 1848, when Louis Pasteur proposed that it had a molecular basis originating from some form of dissymmetry, with the term chirality being coined by Lord Kelvin a year later. The origin of chirality itself was finally determined in 1874, when Jacobus Henricus van't Hoff and Joseph Le Bel independently proposed the tetrahedral geometry of carbon (models up until this time had been 2D) and theorised that the arrangement of groups around this tetrahedron could dictate the optical activity of the resulting compound.

In 1894 Hermann Emil Fischer outlined the concept of asymmetric induction; in which he correctly ascribed selective the formation of D-glucose by plants to be due to the influence of optically active substances within chlorophyll. Fischer also successfully performed what would now be regarded as the first example of enantioselective synthesis, by enantioselectively elongating sugars via a process which would eventually become the Kiliani–Fischer synthesis.

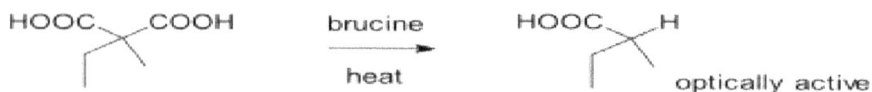

Marckwald Asymmetric Synthesis :

However, the first enantioselective synthesis is often attributed to Willy Marckwald; who in 1904 described the enantioselective decarboxylation of the malonic acid 2-ethyl-2-methylmalonic acid, mediated by brucine. The reason for this is historical, as at the time enantioselective synthesis was seen in terms of vitalism. This argued that natural and artificial compounds were fundamentally different and that chirality could only exist in natural compounds. Unlike Fischer, Marckwald had performed an enantioselective reaction upon an achiral and artificial starting material, making this the first successful enantioselective reaction by the standards of the time.

Much of this early work was published in German, however contemporary English accounts can be found in the papers of Alexander McKenzie.

Early work (1905 - 1965)

The development of enantioselective synthesis was initially slow, largely due to the limited range of techniques available for their separation and analysis. Diastereomers posses different physical properties, allowing separation by conventional means, however at the time enantiomers could only be separated by spontaneous resolution (where enantiomers separate upon crystallisation) or kinetic resolution (where one enantiomer is selectively destroyed). The only tool for analysing enantiomers was optical activity using a polarimeter, a method which provides no structural data.

It was not until the 1950's that major progress really began. Driven in part by chemists such as R. B. Woodward and Vladimir Prelog had derived this part by development of new techniques. The first of these was X-ray Crystallography, which was used to determine the absolute configuration of an organic compound by Johannes Bijvoet in 1951. Chiral chromatography was introduced a year later by Dalgliesh, who used paper chromatography to separate chiral amino acids. Although Dalgliesh was not the first to observe such separations, he correctly attributed the separation of enantiomers to differential retention by the chiral cellulose. This was expanded upon in 1960, when Klem and Reed first reported the use of chirally-modified silica gel for chiral HPLC chromatographic separation. Modern Age (1965 - Present Day)

The Cahn–Ingold–Prelog priority rules (often abbreviated as the CIP system) were first published in 1966; allowing enantiomers to be more easily and accurately described. The same year saw first successful enantiomeric separation by gas chromatography an important development as the technology was in common use at the time.

CHAPTER – II

CHIRALITY IN SCIENCE AND LIFE

Introduction

2.1   What is Chirality ?

Imagine that you had a left hand transplanted in the place of your right hand. Now try and shake hands with a friend. You're in trouble! Whichever way the left hand was transplanted, your fingers would seem to be inverted. The moral is your left hand and your right hand are different. On the other hand, your left hand and your right foot are also different. However, you may feel there exists a relationship between your left hand and your right hand that doesn't exist between your left hand and your right foot. Here's a hint. Put your left hand up to a mirror. The image that appears in the mirror is a righthand. The answer: left and right hands are mirror images.

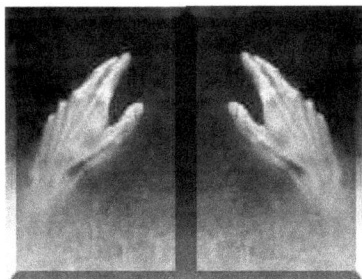

Not all objects behave like a pair of hands. The common water glass, for example is identical to its image in the mirror. The underlying issue is the identity of an object and its reflection. Any object that is different from its reflection is said to be chiral. Otherwise, it is called achiral. Your left hand and your right hand are chiral, whereas the water glass is achiral. Basically, chirality is "handedness," that is the existence of left/right opposition. Even if you had never heard this word, you do know the phenomenon. Otherwise, you would not be able to distinguish left from right.

2.2   Chirality in basic science

In Mathematical Science ,a figure is chiral if it cannot be mapped to its mirror image by rotations and translations alone. For example, a right shoe is different from a left shoe and clockwise is different from anti-clockwise.

A chiral object and its mirror image are said to be enantiomorphs. The helix (and by extension a spun string, a screw, a propeller, etc.) and Mobius strip are chiral two-dimensional objects in three dimensional ambient space. The J, L, S and Z shaped tetrominoes of the popular video game. Tetris also exhibit chirality, but only in a two-dimensional space.

Many other familiar objects exhibit the same chiral symmetry of the human body such as gloves, glasses (where two lenses differ in prescription) and shoes. A similar notion of chirality is considered in knot theory as explained below.

Some chiral three-dimensional objects such as the helix can be assigned a right or left handedness according to the right-hand rule.

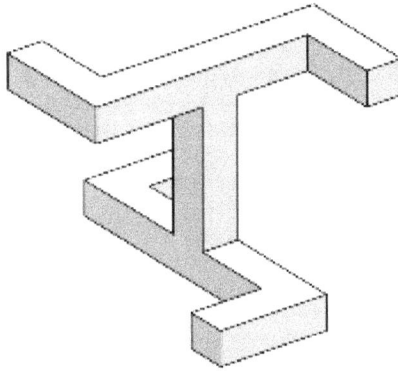

Achiral 3D object without central symmetry or a plane of symmetry

Knot theory: A knot is called achiral if it can be continuously deformed into its mirror image, otherwise it is called chiral. For example the unknot and the figure-eight knot are achiral, whereas the trefoil knot is chiral.

In physics, chirality may be found in the spin of a particle which may be used to define a handedness (chirality) for that particle. A symmetry transformation between the two is called parity. Invariance under parity by a Dirac fermion is called chiral symmetry.

Electro-magnetism

Electromagnetic wave propagation as handedness is wave polarization and described in terms of helicity (occurs as a helix). Polarization of an electro magnetic wave is the property that describes the orientation i.e. the time-varying, direction (vector), and amplitude of the electric field vector.

In Chemistry

A chiral molecule is a type of molecule that has a non-superimposable mirror image. The feature that is most often the cause of chirality in molecules is the presence of an asymmetric carbon atom. The term chiral in general is used to describe an object that is non superposable on its mirror image. chirality usually refers to molecules. Two mirror images of a chiral molecule are called enantiomers or optical isomers. Pairs of enantiomers are often designated as "right-" and "left-handed."

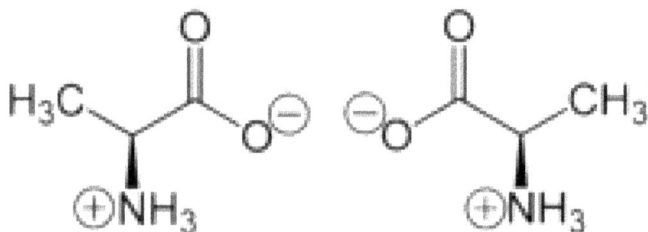

(S)-Alanine (left) and (R)-alanine (right) in zwitter ionic form at neutral pH

Molecular chirality is of interest because of its application to stereochemistry in inorganic chemistry, organic chemistry, physical chemistry, biochemistry and supra molecular chemistry.

In Biology

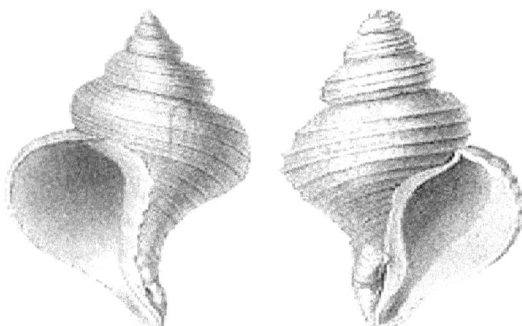

Shells of two different species of sea snail: on the left is the normally sinistral (left-handed) shell of Neptunea angulata, on the right is the normally dextral (right-handed) shell of Neptunea despecta

In anatomy, chirality is found in the imperfect mirror-image symmetry of many kinds of animal bodies. Organisms such as gastropods exhibit chirality in their coiled shells resulting in an asymmetrical appearance. Over 90% of gastropod species have dextral (right-handed) shells in their coiling but a small minority of species and genera are virtually always sinistral (left-handed). A very few species (for example Amphidromus perversus) show an equal mixture of dextral and sinistral individuals.

In humans, chirality (also referred to as handedness or laterality) is an attribute of humans defined by their unequal distribution of fine motor skill between the left and right hands. An individual who is more dexterous with the right hand is called right-handed and one who is more skilled with the left is said to be left-handed. Chirality is also seen in the study of facial asymmetry.

In flat fish, the Summer flounder or fluke are left-eyed, while halibut are right-eyed.

2.3 To whom is this site proposed?

You could simply be curious about symmetry issues. No background in science is needed to browse through the "beginners" section (still to be written!). This site was tentatively conceived to provide you information, hopefully without boring you with scientific jargon. If you feel that this site is a failure in that respect, you are encouraged to complain here.

Besides, research results will also be presented. Chirality is an old theme in many branches of chemistry and biology with overwhelming practical importance and physicists have long paid interest in symmetry matters. As a result, there is a wealth of published articles dealing with chirality in the physical sciences. However, this site is primarily focused on recent theoretical progress in a quite unexpected direction: the description of chirality.

2.4   Started Awarness for Life through chirality

Thalidomide is a sedative drug that was prescribed to pregnant women, from 1957 into the early 60's. It was present in at least 46 countries under different brand names. "When taken during the first trimester of pregnancy, Thalidomide prevented the proper growth of the foetus, resulting in horrific birth defects in thousands of children around the world". Why? The Thalidomide molecule is chiral. There are left and right-handed Thalidomides, just as there are left and right hands. The drug that was marketed was a 50/50 mixture. One of the molecules, say the left one was a sedative, whereas the right one was found later to cause foetal abnormalities. "The tragedy is claimed to have been entirely avoidable had the physiological properties of the individual thalidomide molecules been tested prior to commercialization."

Aspartame is a sweetening agent that is more than a hundred times sweeter than sucrose. And yet, the mirror image molecule is bitter. "(S)-carvone possesses the odor perception of caraway while the mirror image molecule (R)-carvone has a spearmint odor."

These examples are just the tip of the iceberg. DNA, proteins, amino acids, sugars are all chiral. Mirror image amino acids are called L-

and D-amino acids. Human proteins are exclusively built from L-amino acids. The origin of this fundamental dissymmetry is still mysterious. When interacting, molecules recognize each other just as your right hand distinguishes another right hand from a left when you shake hands. This is why mirror image molecules, like mirror image Thalidomides, so often have radically different fates in our bodies.

Many of the building blocks of biological systems, such as sugars and amino acids are produced exclusively as one enantiomer. As a result of this living systems possess a high degree of chemical chirality and will often react differently with the various enantiomers of a given compound. Examples of this selectivity include:

Flavour : the artificial sweetener aspartame has two enantiomers. L-aspartame tastes sweet yet D-aspartame is tasteless

Odor : R-(–)-carvone smells like spearmint yet S-(+)-carvone smells like caraway.

Drug effectiveness: the antidepressant drug Citalopram is sold as a racemic mixture. However studies have shown that only the (S)-(+) enantiomer is responsible for the drugs beneficial effects.

Drug safety: D-penicillamine is used in chelation therapy and for the treatment of rheumatoid arthritis. However L-penicillamine is toxic as it inhibits the action of pyridoxine.

CHAPTER – III

INTRODUCTION OF ISOMERS IN CHIRAL MOLECULE

3.1 Origin of life: The Chirality Problem

Many important molecules required for life exist in two forms. These two forms are non-superimposable mirror images of each other, i.e.: they are related like our left and right hands. Hence this property is called chirality from the Greek word for hand. The two forms are called enantiomers (from the Greek word for opposite) or optical isomers because they rotate plane-polarised light either to the right or to the left.

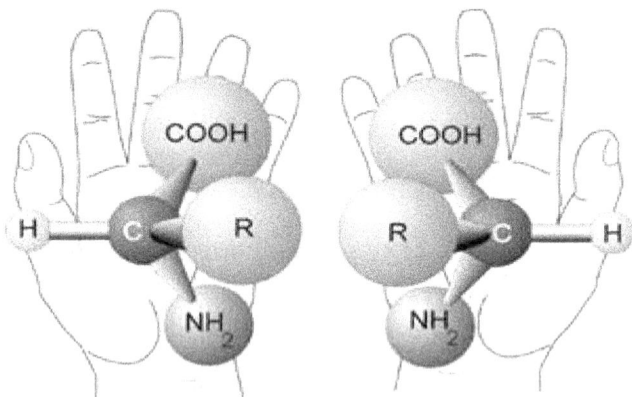

Diagram of chirality

Whether or not a molecule or crystal is chiral is determined by its symmetry. A molecule is achiral (non-chiral) if and only if it has an axis of improper rotation, that is an n-fold rotation (rotation by $360°/n$) followed by a reflection in the plane perpendicular to this axis maps the molecule on to itself. Thus a molecule is chiral if and only if it lacks such an axis. Because chiral molecules lack this type of symmetry, they are called dissymmetric. They are not necessarily asymmetric (i.e. without symmetry) because they can have other types of symmetry. However, all amino acids (except glycine) and many sugars are indeed asymmetric as well as dissymmetric.

Nearly all biological polymers must be homochiral (all its component monomers having the same handedness. Another term used is optically pure or 100 % optically active) to function. All amino acids in proteins are 'left-handed' while all sugars in DNA and RNA and in the metabolic pathways are 'right-handed'.

A 50/50 mixture of left- and right-handed forms is called a racemate or racemic mixture. Racemic polypeptides could not form the specific shapes required for enzymes because they would have the side chains sticking out randomly. Also, a wrong-handed amino acid disrupts the stabilizing α-helix in proteins. DNA could not be stabilized in a helix if even a single wrong-handed monomer were present, so it could not form long chains. This means it could not store much information, so it could not support life.

3.2   Introduction for enantiomers

Normally, the two enantiomers of a molecule behave identically to each other. For example, they will migrate with identical R_f in thin layer chromatography and have identical retention time in HPLC. Their NMR and IR spectra are identical. However, enantiomers behave differently in the presence of other chiral molecules or objects. For example, enantiomers do not migrate identically on chiral chromatographic media such as quartz or standard media that have been chirally modified. The NMR spectra of enantiomers are affected differently by single-enantiomer chiral additives such as EuFOD.

Chiral compounds rotate plane polarized light. Each enantiomer will rotate the light in a different sense, clockwise or counterclockwise. Molecules that do this are said to be optically active.

Characteristically, different enantiomers of chiral compounds often taste and smell differently and have different effects as drugs. These effects reflect the chirality inherent in biological systems.

One chiral 'object' that interacts differently with the two enantiomers of a chiral compound is circularly polarised light: An enantiomer will absorb left- and right-circularly polarised light to differing degrees. This is the basis of circular dichroism (CD) spectroscopy. Usually the difference in absorptivity is relatively small (parts per thousand). CD spectroscopy is a powerful analytical technique for investigating the secondary structure of proteins and for determining the absolute configurations of chiral compounds, in particular transition metal complexes. CD spectroscopy is replacing polarimetry as a method for characterising chiral compounds, although the latter is still popular with sugar chemists.

Many biologically active molecules are chiral, including the naturally occurring amino acids (the building blocks of proteins) and sugars. In biological systems, most of these compounds are of the same chirality: most amino acids are l and sugars are d. Typical naturally occurring proteins, made of l amino acids are known as left-handed proteins, where as d amino acids produce right-handed proteins.

The origin of this homochirality in biology is the subject of much debate. Most scientists believe that Earth life's "choice" of chirality was purely random and that if carbon-based life forms exist elsewhere in the universe, their chemistry could theoretically have opposite chirality. However, there is some suggestion that early amino acids could have formed in comet dust. In this case, circularly polarised radiation (which makes up 17% of stellar radiation) could have caused the selective destruction of one chirality of amino acids, leading to a selection bias which ultimately resulted in all life on Earth being homochiral.

Enzymes, which are chiral often distinguish between the two en-antiomers of a chiral substrate. Imagine an enzyme as having a glove-like cavity that binds a substrate. If this glove is right-handed then one enantiomer will fit inside and be bound whereas the other enantiomer will have a poor fit and is unlikely to bind.

d-form amino acids tend to taste sweet, whereas l-forms are usually tasteless. Spearmint leaves and caraway seeds respectively, contain R-(–)-carvone and S-(+)-carvone-enantiomers of carvone. These smell different to most people because our olfactory receptors also contain chiral molecules that behave differently in the presence of different enantiomers.

Chirality is important in context of ordered phases as well, for example the addition of a small amount of an optically active molecule to a nematic phase (a phase that has long range orientational order of molecules) transforms that phase to a chiral nematic phase (or cholesteric phase). Chirality in context of such phases in polymeric fluids has also been studied in this context.

3.3 d-Amino Acid Natural Abundance

The relative abundances of each of the different d-isomers of several amino acids have recently been quantified by collecting experimentally reported data from the proteome across all organisms in the Swiss-Prot database. The d-isomers observed experimentally were found to occur very rarely as shown in the following table in the database of protein sequences containing over 187 million amino acids

CHAPTER – IV

HOMOCHIRALITY IN MOLECULE OF LIFE

4.1 Homochirality: A Major Problem for Origin of Life Theories

Introduction

The primary molecules of life exist in one specific form even though ordinary chemical reactions produce equal amount of both mirror forms. Living organisms contain complex enzymes that specifically produce molecules of only one form. This article discusses the problems associated the generation of specific enantiomers of amino acids or sugar within naturalistic origin of life theories. Since the jargon is necessarily specific to the topic, you may want to review the concepts & terminology of introductory molecular biology in order to understand the concepts.

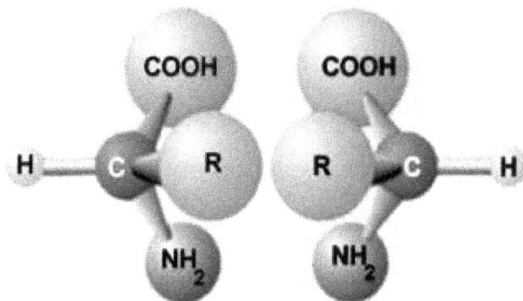

All living organisms are based upon certain "mirror" forms of amino acids and sugars. Although normal chemical reactions produce right and left mirrors in equal amounts ("racemic" mixtures), life uses specialized molecular machinery to produce only right handed forms of sugars and left handed forms of amino acids (called "enantiomers"). Because these chemicals exist in only one form, they are referred to as being "homochiral." Origin of life theories must explain how chemistry could produce the proper mirrored building blocks to support the generation of the first self-replicating life form.

4.2   Why homo chirality is important ?

Origin of life theories often ignore the homochirality problem, even though the question is critical to the origin of life. Both DNA and RNA are incapable of complementary pair bonding in the absence of being homochiral. Practically, this means that racemic DNA or RNA cannot replicate. Because of this problem, organic chemist William Bonner has dismissed the point of view that homochirality in nucleic acids and/or amino acids did not precede the origin of life. Because of the importance of the question, Bonner spent considerable effort looking for a solution but admitted terrestrial mechanisms for homochirality and trying to experimentally investigate them and didn't find any supporting evidence." He added that "Terrestrial explanations are impotent or nonviable."

Miller's Electrical Discharge Unit :

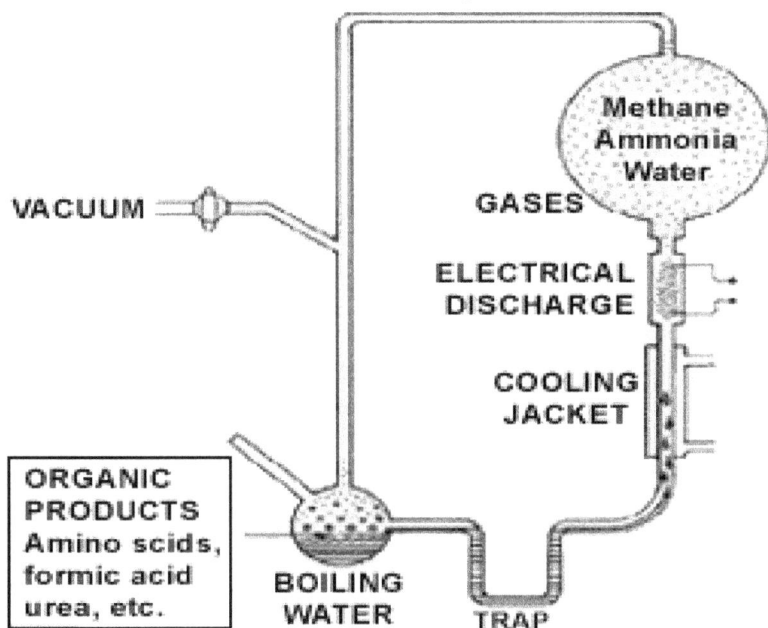

4.3 Amino Acids

Researcher Stanley Miller demonstrated in 1953 that mixtures of reducing gases thought to be present in the primordial earth when subjected to electrical discharges produced many organic compounds, including several amino acids.

Years later in 1969 a meteor that landed in Murchison, Australia, was shown to contain the same organic compounds and amino acids, in roughly the same proportion as those generated through the Miller experiments. An initial examination of the chirality of the amino acids revealed either no chiral excesses or L-amino acid excesses that were probably the result of contamination by terrestrial sources. However, in 1997, a study examined chiral proportions of non terrestrial amino acids. The results showed a 7% and 9% L- enantio-meric excess within extracts of the meteorite. Although statistically significant, such as small excess of L-amino acids would not solve the problem of 100% L-enantiomeric excess required by earth's life forms.

The carbonaceous meteorite GRA 95229 was discovered in 1824 in Antarctica and appears to be relatively free of terrestrial contamination. Analysis of the organic materials revealed the presence of several amino acids and aldehydes with L-enantiomer excesses up to 14%. Again, such excesses are not enough to account for the origin of homochirality. The most pristine carbonaceous chondrite examined to date is the Tagish Lake Meteorite, which was observed to fall on a frozen lake in Canada during Winter, 2000 and was collected to minimize human contamination. Amino acids in the meteorite were observed only in parts-per-billion concentrations and no L-enantiomeric excess was found.

An attempt to be explain the origin of homochirality in amino acids has been made by invoking some rather complex chemistry. Researchers have shown that transfer RNA (tRNA), the molecule responsible for binding to amino acids during protein synthesis, does not selectively bind to L-amino acids in solution. So, it has become apparent that the prebiotic soup hypothesis will not explain the origin of homochirality. However, researchers thought that if RNA were bound to a solid substrate, the selectivity of RNA might be constrained, since the molecular configuration would be more limited. So, researchers have hypothesized that homochiral RNA bound to polar mineral surfaces would alter stereospecificity. In fact, this technique has produced up to a 35-60% enantiomeric excess of L-amino acids but not nearly enough to produce any reasonably-sized protein.

In another strategy, researchers isolated the amino acid binding site of several tRNAs and modified them to make an "RNA minihelix" that demonstrated a four-fold enantiomeric selection. Although impressive compared to previous attempts, such a system would still produce a 20% error rate, which would prevent the formation of anything larger than a small peptide. In addition, the system assumes that the problem of RNA homochirality had already been solved. In addition, the author assumed that some primitive life form would have come up with the minihelix design instead of stealing it from current life forms, which is what the scientists did.

Another means by which L-amino acids could be selected is through a specific molecule that prevents or removes binding of D-amino acids to tRNAs. In most living systems, specific molecules are used to discriminate between structurally similar amino acids. In general, the tRNA itself cannot discriminate and allows the smaller amino acid to bind. However, another molecule will remove the smaller intruder amino acid, preserving the accuracy of the genetic code. It turns out that in some species of Archaea (ancient, bacteria-like, single celled organisms), the molecule that corrects for improper amino acid charging can also distinguish between certain L- and D-amino acids. It is possible that such a system could select for homochiral amino acids. However, the requirement for 20 specifically-designed accessory molecules (one for each amino acids) would add a level of design that would not be expected in primitive early life forms.

A recent study has found differential separation of racemic mixtures of the amino acid proline when dissolved at high concentrations (25-100 mM) in pure DMSO. Aldol condensation reactions produce amino acids under these conditions with an enantiomeric excess from 46-99%. Although this seems impressive, there would have been no prebiotic source of high concentrations of purified amino acids nor DMSO solvent.

Peptides

Attempts to produce amino acids have been plagued with problems involving unreactive by products. In particular, the formation of peptides under primordial conditions have resulted in the formation of large amounts of unreactive diketo-piperazines. This problem can be circumvented by incubating purified L-amino acids (obviously not available on the primordial earth) in a slurry of (Ni, Fe) at boiling temperatures under alkaline conditions (similar to those observed in undersea hydrothermal vents). The problems with such a system was that, although the system produced very short peptides, the process itself resulted in racemization of the peptides. In addition, under these conditions, these peptides hydrolyzed rapidly (reversal of the process). Ultimately, because these problems, the process produced low amounts of usable peptides.

Ribose

The sugar ribose forms the backbone of RNA (and its related sugar, deoxyribose makes up the backbone of DNA). Researchers have shown that enantiomeric excess of sugars can be produced by using chiral amino acid catalysts. However, even a 100% homochiral amino acid increased the enantiomeric excess by only 10%. Subtle chiral excesses of amino acids (like those found in meteorites) negligible produced enantiomeric excess in synthesized sugars.

4.4   Homochirality in extra-terrestrial sources

In 1997, scientists hypothesized that the enantiomeric excess of L-amino acids in extraterrestrial sources could be due to circular polarization of synchrotron radiation in neutron stars, which selectively destroys the opposite handed enantiomer. According to this theory, a neutron star was originally present near the interstellar molecular cloud from which the Solar System formed, resulting in excess L-amino acids and/or their precursors. However, over time, such radiation would destroy all amino acids even the enantiomer that is destroyed at a lower rate. In addition, over the entire spectrum of circularly polarized radiation, the susceptibility of specific enantiomers to differential photochemical degradation sum to zero. So, any preferential degradation of an amino acid would require some means of providing monochromatic circularly polarized radiation, which is extremely unlikely. There are even more problems with circularly polarized radiation theories. Although, such sources of radiation are theoretically possible, the reality is that none have yet been found. For example, the Crab Nebula has been proposed as a possible source of synchrotron radiation but detection has shown a maximum amount of 0.03%. What has been detected is a level of 0.05% at a wavelength of 1415 MHz. However, this wavelength is one million times longer than the wavelengths that actually produce an effect. So, to date, there is no evidence that natural sources of circularly polarized radiation of the proper wavelength actually exist.

Despite these problems, researchers have found that phthalic acid crystals differentially scatter circularly polarized light, making it possible that such crystals might be involved somehow in the synthesis of homochiral amino acids. Now, where did the universe put those phthalic acid crystals?

4.4   Remark On Homochirality:

The origin of homochirality is extremely important in origin of life research, since non-optically pure mixtures of amino acids or sugars cannot be used to make RNA, DNA and proteins, the building blocks of all living organisms. There is no terrestrial or extraterrestrial explanation that describes how homochirality could have arisen through completely naturalistic processes. Processes that can enhance the enantiomeric excess of appropriate amino acid or nucleic acid building blocks produce only modest increases in the percentage of those products, while requiring unrealistic, laboratory conditions, which could not have been present on the primordial earth. Coupled with the inability of unaided chemistry to even produce some of the required molecular building blocks of life, a completely naturalistic origin of life seems extremely unlikely. Such insurmountable problems led us, as an undergraduate biology major at USC, to the conclusion that at minimum, there must be a Creator God who designed the first life form, prompting me to go from atheism to deism.

CHAPTER – V

TOOL FOR IMPORTANT PRACTICAL ASPECTS

5.1   Industrial Aspects for Chirality

We have come through from the chirality in daily life with some examples. But this event throws a light in industrial point of view much more.

It is mean that 'Chirality' word is referred by Physical event where as 'Enatiomer' word is for natural or synthetic compounds. We understand always compounds are related to chemical industry or other industry

Although converting a molecule from one enantiomer to the other seems like only a small change in the structure, it can provide a significant impact on the way the molecule interacts with its surroundings and especially other chiral compounds. Many of the molecules that are important in nature are chiral, these include proteins (and their constituent amino acids), which control most processes within biological systems and the nucleic acids DNA and RNA which are responsible for holding the information necessary for proteins to be synthesised.

For this reason, if a chiral compound interacts with a protein to induce a specific response in a biological organism, it is likely that its enantiomer will either not interact or produce a completely different response. Some of these differences can be quite startling, for example limonene contains a chiral carbon atom. One enantiomer produces the smell of oranges whereas the other gives rise to the smell of lemons.

Understanding chirality is extremely important in the preparation of therapeutic drugs. For example, one enantiomer of penicillamine is a potent anti-arthritic agent whereas the other enantiomer is highly toxic. Perhaps the most startling example of the difference in activity between enantiomers is Thalidomide. This drug was seen as a panacea for the treatment of morning sickness in pregnant women and indeed one enantiomer reliably has this effect. The other enantiomer, unfortunately, has been associated with the well-characterised birth defects that arose from use of Thalidomide.

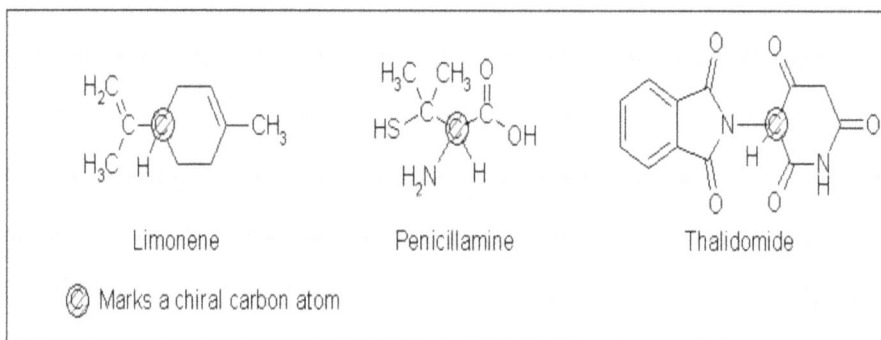

Limonene   Penicillamine   Thalidomide

Ⓒ Marks a chiral carbon atom

One further difference between enantiomers is the way that they interact with light. Light is an electromagnetic radiation. This means that it consists of electronic and magnetic components. The electronic components of light interact with electrons, such as bonds within a molecule. Changing the arrangement of the bonds changes the way that light interacts with the molecule. This difference in chiral molecules only becomes apparent when polarised light is shone through a solution of the molecule. In polarised light, all the electronic components are aligned. As the polarised light passes through the solution of a chiral compound the polarised light is twisted with the plane of polarisation being rotated.

Two enantiomers rotate the plane of polarised light by equal amounts but in opposite directions. For this reason, stereoisomerism is also sometimes referred to as optical isomerism.

5.2 Need & tool for Chiral compounds:

Several compounds in routine use need of health consciousness & environmental awareness bearing chiral centre. The compounds having two chiral centres (racemates) can no longer be used. Because Chirally pure compounds(Enantiomers) may exhibit very different type of biological activity, both of which may have beneficial ,one of them is beneficial and other may be undesirable, harmful or no effect.

How to separate both enatiomer ? This was a big question in earlier days of 1990's.The industrial transformation requires a greater precision in terms of not only one isomeric selectivity but high stereoselectivity with a specific properties, activity and functions.

Today, scientists are engaged in the design & synthesis of molecular substances that will display specific properties, activities and functions after Pasteur's landmark revelation of the molecule origins of optical activity.

Although very often enantioselectivities of 90-95% have been achieved for a variety of transformations in the research laboratory. This is usually not sufficient for industrial needs, particularly when it comes to pharmaceutical intermediate & products also in agrochemicals, favours, fragrances, nutrients, cosmetics & Vitamins.

One survey showed that among one third part of the drug sale in market covered before some year ago by a chiral drugs.

The certification of pharmaceutical product requires compound to have an enantiomeric purity of 99% or above it.

Chiral Synthesis is emerged as most preffered and powerful choice to obtain enantiopure compounds.

"How to make good asymmetric synthesis perfect" must therefore be the main concern for everybody, not only industrial Chemist.

Enantioselective synthesis, also called chiral synthesis, asymmetric synthesis or stereo selective synthesis, is defined by IUPAC as: a chemical reaction (or reaction sequence) in which one or more new elements of chirality are formed in a substrate molecule and which produces the stereoisomeric (enantiomeric or diastereoisomeric) products in unequal amounts. Put more simply: it is the formation of a compound as single enantiomer or diastereomer.

Enantioselective synthesis is a key process in modern chemistry and is particularly important in the field of pharmaceuticals, as the different enantiomers or diastereomers of a molecule often have different biological activity.

5.3 Approaches :

Enantioselective catalysis

In general, enantioselective catalysis (known traditionally as asymmetric catalysis) refers to the use of chiral coordination complexes as catalysts. It is very commonly encountered, as it is effective for a broader range of transformations than any other method of enantioselective synthesis. The catalysts are typically rendered chiral by using chiral ligands, however it is also possible to generate chiral at metal complexes using simpler achiral ligands. Most enantioselective catalysts are effective at low concentrations making them well suited to industrial scale synthesis; as even exotic and expensive catalysts can be used affordably. Perhaps the most versatile example of enantioselective synthesis is asymmetric hydrogenation, which is able to reduce a wide variety of functional groups.

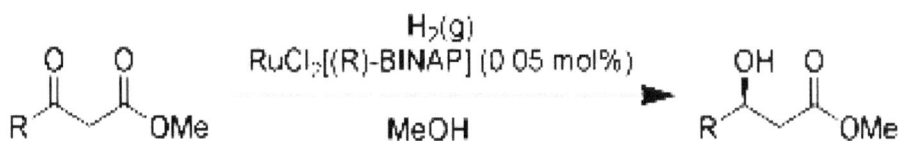

With only 75 natural metals in existence (and not all of these showing extensive catalytic activities) the design of new catalysts is very much dominated by the development of new classes of ligands. However, certain ligands have been found to be effective in a wide range of reactions, many of which proceed via different mechanisms. These are often referred to as 'privileged ligands' and include examples such as BINOL, Salen and BOX.

Chiral auxiliaries

A chiral auxiliary is an organic compound which couples to the starting material to form new compound which can then undergo enantioselective reactions via intramolecular asymmetric induction. At the end of the reaction the auxiliary is removed under conditions that will not cause racemization of the product. It is typically then recovered for future use.

$$\text{H–X}_c$$

chiral
auxiliary

auxiliary
recycle

$$\underset{\substack{\text{prochiral}\\\text{substrate}}}{\text{HO}\overset{\text{O}}{\diagup}\text{R}} \longrightarrow \underset{\substack{\text{diastereoselective}\\\text{transformation}}}{X_c\overset{\text{O}}{\diagup}\text{R}} \longrightarrow X_c\overset{\text{O}}{\diagup}\underset{E}{\text{R}} \longrightarrow \underset{\text{chiral product}}{\text{HO}\overset{\text{O}}{\diagup}\underset{E}{\text{R}}}$$

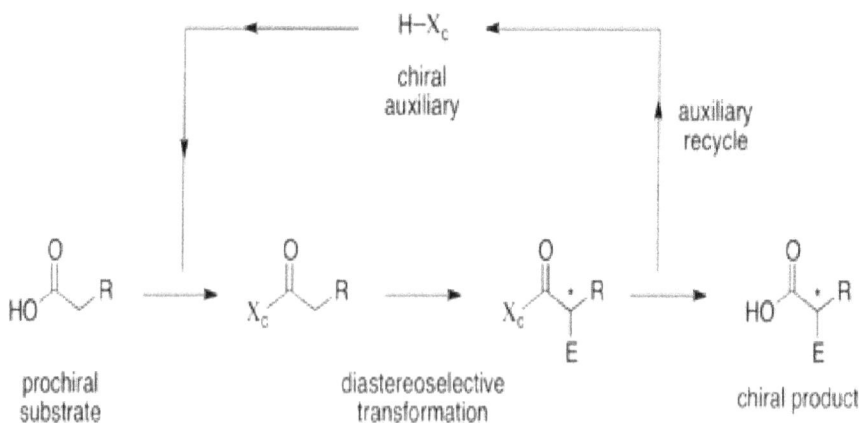

Chiral auxiliaries must be used in stoichiometric amounts to be effective and require additional synthetic steps to append and remove the auxiliary. However in some cases the only available stereoselective methodology relies on chiral auxiliaries and these reactions tend to be versatile and very well-studied, allowing the most time-efficient access to enantiomerically pure products. Additionally, the products of auxiliary-directed reactions are diastereomers which enables their facile separation by methods such as column chromatography or crystallization.

Biocatalysis

Biocatalysis makes use of biological compounds, ranging from isolated enzymes to living cells, to perform chemical transformations. The advantages of these reagents include very high eel's and reagent specificity, as well as mild operating conditions and low environmental impact. Biocatalysts are more commonly used in industry than in academic research; for example in the production of statins. The high reagent specificity can be a problem however; as it often requires that a wide range of biocatalysts be screened before an effective reagent is found.

Enantioselective organocatalysis

Organocatalysis refers to a form of catalysis, where the rate of a chemical reaction is increased by an organic compound consisting of carbon, hydrogen, sulfur and other non-metal elements. When the organocatalyst is chiral enantioselective synthesis can be achieved; for example a number of carbon-carbon bond forming reactions become enantioselective in the presence of proline with the aldol reaction being a prime example. Organocatalysis often employs natural compounds and secondary amines as chiral catalysts; these are inexpensive and environmentally friendly, as no metals are involved.

Chiral pool synthesis

Chiral pool synthesis is one of the simplest approaches for enantioselective synthesis, as it does not involve asymmetric induction. Instead a chiral starting material is manipulated through successive reactions, using achiral reagents to obtain the desired target molecule. This can meet the criteria for enantioselective synthesis when a new chiral species is created such as in an SN^2 reaction. Chiral pool synthesis is especially attractive for target molecules having similar chirality to a relatively inexpensive naturally occurring building-block such as a sugar or amino acid. However, the number of possible reactions the molecule can undergo is restricted and tortuous synthetic routes may be required (e.g. Oseltamivir total synthesis).

$$\text{Nu} \quad \overset{X}{\underset{Z}{\overset{|}{\underset{|}{C}}}}\text{---}Y \quad L \quad \longrightarrow \quad \left[\text{Nu}\text{-----}\overset{X}{\underset{\underset{Z}{Y}}{\overset{\delta+}{C}}}\text{----}L^{\delta-} \right]^{\ddagger} \quad \longrightarrow \quad \text{Nu} \quad \overset{X}{\underset{Z}{\overset{|}{\underset{|}{C}}}}\text{'''}Y \quad + \quad L$$

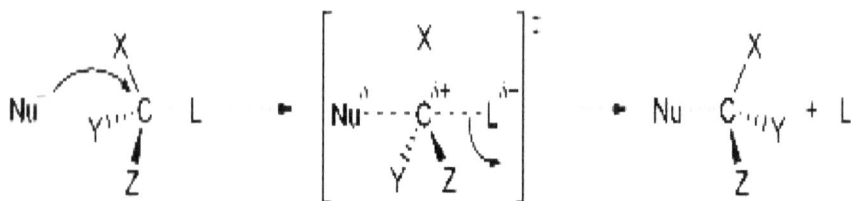

This approach also requires a stoichiometric amount of the enantiopure starting material, which can be expensive if it is not naturally occurring.

5.4   Alternative approaches:

Apart from enantioselective synthesis, chiraly pure materials can be obtained by chiral resolution. This involves the isolation of one enantiomer from a racemic mixture by any of a number of methods. Where the cost in time and money of making such racemic mixtures is low, or if both enantiomers may find use, this approach may remain cost-effective.

Separation and Analysis of Enantiomers

The two enantiomers of a molecule possess the same physical properties (e.g. melting point, boiling point, polarity etc.) and so behave identically to each other. As a result they will migrate with an identical R_f in thin layer chromatography and have identical retention times in HPLC and GC. Their NMR and IR spectra are identical.

This can make it very difficult to determine whether a process has produced a single enantiomer (and crucially which enantiomer it is) as well as making it hard to separate enantiomers from a reaction which has not been 100% enantioselective. Fortunately, enantiomers behave differently in the presence of other chiral materials and this can be exploited to allow their separation and analysis.

Enantiomers do not migrate identically on chiral chromatographic media, such as quartz or standard media that has been chirally modified. This forms the basis of chiral column chromatography which can be used on a small scale to allow analysis via GC and HPLC or on a large scale to separate chirally impure materials. However this process can require large amount of chiral packing material which can be expensive. A common alternative is to use a chiral derivatizing agent to convert the enantiomers into a dia-stereomers, in much the same way as chiral auxiliaries. These have different physical properties and hence can be separated and analysed using conventional methods. Special chiral derivitizing agents know as 'chiral resolution agents' are used in the NMR spectroscopy of stereoi-somers, these typically involve coordination to chiral europium complexes such as $Eu(fod)_3$ and $Eu(hfc)_3$.

The enantiomeric excess of a substance can also be determined using certain optical methods. The oldest method for doing this is to use a polarimeter to compare the level of optical rotation in the product against a 'standard' of known composition. It is also possible to perform ultraviolet-visible spectroscopy of stereoisomers by exploiting the Cotton effect.

One of the most accurate ways of determining the chirality of compound is to determine its absolute configuration by X-ray Crystallography. However this is a labour intensive process which requires that a suitable single crystal be grown.

CHAPTER – VI

CHIRAL COMPOUND IN IMPORTANT INDUSTRY

6.1 Importance of Chirality

Catalytic asymmetric synthesis

Three researchers share in 2001 year's Nobel Prize in Chemistry; Dr .William S. Knowles, who had been working at Monsanto Company, St Louis, USA; Professor Ryoji Noyori, Nagoya University, Chikusa, Nagoya, Japan and Professor K. Barry Sharpless, The Scripps Research Institute, La Jolla, California, USA. The Royal Swedish Academy of Sciences rewards the three chemists for: "their development of catalytic asymmetric synthesis". Knowles and Noyori receive half the Prize for: "their work on chirally catalysed hydrogenation reactions" and Sharpless is rewarded with the other half of the Prize for: " his work on chirally catalyzed oxidation reactions". The discoveries made by the three organic chemists have had a very great impact on academic research and the development of new drugs and materials and are used in many industrial syntheses of drugs and other biologically active compounds. Below is given a background and description of their discoveries.

Chiral molecules

Year of 2001 Nobel Prize in Chemistry concerns the development of chiral transition metal catalysts for stereoselective hydrogenations and oxidations - two important classes of synthetic reactions. Through the Laureates' work a myriad of useful chiral compounds have become accessible. Many of the compounds associated with living organisms are chiral, for example DNA, enzymes, antibodies and hormones. Therefore enantiomers of compounds may have distinctly different biological activity. Thus the enantiomers of limonene, both formed naturally, smell differently - one of the enantiomers (S)-limonene smells of lemons, while the mirror image compound (R)-limonene smells of oranges (R)-Limonene smells of oranges and (S)-limonene smells of lemons. We distinguish between these enantiomers because our nasal receptors are also made up of chiral molecules that recognise the difference. Insects use chiral chemical messengers (pheromones) as sex attractants and chemists have discovered that one of the enantiomers of the insect pheromone, olean, attracts male fruit flies, while its mirror image operates on the female of the species.

Thus biology is very sensitive to chirality and the activity of drugs also depends on which enantiomer is used. Most drugs consist of chiral molecules. And since a drug must match the receptor in the cell, it is often only one of the enantiomers that is of interest.

Pharmaceutical companies nowadays have to make sure that both enantiomers of a drug are tested for their biological activity and toxicity before they are marketed. Obviously, there is a strong demand for to the pure enantiomers.

Catalytic asymmetric syntheses

Industrial companies are concerned about disposing of unwanted compounds and also about the inefficiency and costs involved in the chemical processes. Therefore there is a strong demand for efficient methods for asymmetric syntheses. Thus finding new methods of asymmetric synthesis has in the past 20 or 30 years become a key activity for organic chemists. Ideally, a chiral agent should behave as a catalyst with enzyme-like selectivity. A small amount of material containing the chiral information could generate a large amount of a chiral product. Research has been intensive to develop methods for catalytic asymmetric synthesis i.e. catalytic methods to prepare one of the enantiomers in preference to the other. In a catalytic asymmetric reaction, a chiral catalyst is used to produce large quantities of an optically active compound from a precursor that may be chiral or achiral. In recent years, synthetic chemists have developed numerous catalytic asymmetric syntheses that convert prochiral substrates into chiral products with high enantioselectivity. These developments have had an enormous impact on academic and industral organic syntheses. One single chiral catalyst molecule can direct the stereoselection of millions of chiral product molecules. Such reactions are thus highly productive and economical and when applicable, they make the waste resulting from racemate resolution obsolete. It is researchers in this field who are rewarded with Nobel Prize in Chemistry.

Knowles´pioneer work

In the early sixties it was not known whether catalytic asymmetric hydrogenation was feasible. A breakthrough came in 1968 when Knowles at Monsanto Company, St. Louis showed that a chiral transition metal based catalyst could transfer chirality to a nonchiral substrate resulting in chiral product with one of the enantiomers in excess.Two developments in the mid-sixties offered an attractive approach to making such a catalyst. The first was the discovery by Osborn and Wilkinson of the rhodium complex, $(PPh_3)_3RhCl$, as a soluble hydrogenation catalyst for unhindered olefins. Homogeneous catalysts had been reported earlier but this was the first one that compared in rates with the well-known heterogeneous counterparts.

The other development was the discovery of methods for preparing optically active phosphines by Horner and by Mislow. Knowles' basic strategy was to replace triphenylphosphine in Osborn and Wilkinson's catalyst with the enantiomer of a known chiral phosphine and hydrogenate a prochiral olefine. Knowles soon verified the validity of this thinking by using the known non-racemic methylpropylphenylphosphine (69% of ee of (-)-methylpropylphenylphosphine) and reducing substituted styrenes Knowles's catalytic asymmetric hydrogenation of phenylacrylic acid using a rhodium catalyst containing (-)-methylpropylphenylphosphine (69% ee) gave (+)- hydratropic acid in 15% ee.

A modest enantiomeric excess (ee) was obtained but it was too small to be of any practictical use. However, the result proved that it was in fact possible to achieve catalytic asymmetric hydrogenation. Other researchers (Horner, Kagan, Morrisonand Bosnich) reached similar results shortly afterwards and they have all contributed to opening the doors to a new, exciting and important field for both academic and industrial research.

Noyori's general hydrogenation catalysts

An early example of molecular asymmetric catalysis using homogeneous transition metal complexes - enantioselective cyclopropanation of olefins - was reported in 1966 by Noyori, together with H. Nozaki. However, the stereoselectivity was low. In 1980, Noyori, together with Takaya, discovered an atropisomeric chiral diphosphine, BINAP. Rh(I) complexes of the enantiomers of BINAP are remarkably effective in various kinds of asymmetric catalysis. This includes enantioselective hydrogenation of α- (acylamino) acrylic acids or esters, giving amino acid derivatives and also includes enantioselective isomerization of allylic amines to enamines. The chiral efficiency of BINAP chemistry originates from unique dissymmetric templates created by a transition metal atom or ions and the C_2 chiral diphosphine. Noyori's discovery of the   BINAP Ru(II) complex catalysts was a major advance in stereoselective organic synthesis. The scope of the application of these catalysts is far reaching. These chiral Ruthenium complexes serve as catalyst precursors for the highly enantioselective hydrogenation of a range of α,β- and β,γ-unsaturated carboxylic acids.

Sharpless's chirally catalyzed oxidations

Parallell to the progress in catalytic asymmetric hydrogenations Barry Sharpless has developed chiral catalysts for very important oxidation reactions. The epoxidation reaction discovered in 1980 by Sharpless and Kazuki is a very fine example of a strategy of using a reagent to achieve stereochemical control. Using titanium (IV) tetraisopropoxide, tert-butyl hydroperoxide and an enantiomerically pure dialkyl tartrate, the Sharpless reaction accomplishes the epoxidation of allylic alcohols with excellent stereoselectivity. This powerful reaction is very predictable.When the D-(-)-tartrate ligand (D-(-)-DET) is used in epoxidation, the oxygen atom is delivered to the top face of the olefin (i.e. OH group in lower right hand corner). The L-(+)-tartrate ligand (L-(+)-DET), on the other hand, allows the bottom face of the olefin to be epoxidised. When achiral allylic alcohols are employed, the Sharpless reaction exhibits exceptional enantiofacial selectivity (ca. 100:1) and provides convenient access to synthetically versatile epoxy alcohols.

6.2   Importance of Chirality in Material Science

Electromagnetic wave propagation as chirality is wave polarization and described in terms of helicity (occurs as a helix). Polarization of an electromagnetic wave is the property that describes the orientation, i.e., the time-varying, direction (vector), and amplitude of the electric field vector. For a depiction, see the image to the left.

In the image, it can be seen that polarizations are described in terms of the figures traced as a function of time all along the electric field vector. A representation of the electric field, as a vector, is placed on to a fixed plane in space. The plane is perpendicular to the direction of propagation.

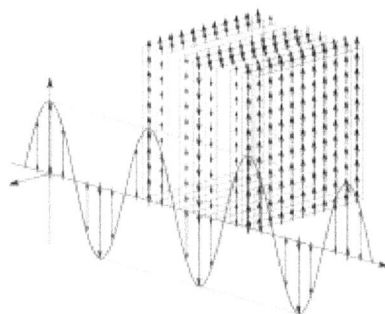

Linearly polarized light. The block of vectors represent how the magnitude and direction of the electric field is constant for an entire plane, which is perpendicular to the direction of travel.

In general, polarization is elliptical and is traced in a clockwise or counterclockwise sense, as viewed in the direction of propagation. If, however, the major and minor axes of the ellipse are equal, then the polarization is said to be circular . If the minor axis of the ellipse is zero, the polarization is said to be linear . Rotation of the electric vector in a clockwise sense is designated right-hand polarization and rotation in a counterclockwise sense is designated left-hand polarization

Mathematically, an elliptically polarized wave may be described as the vector sum of two waves of equal wavelength but unequal amplitude and in quadrature (having their respective electric vectors at right angles and $\pi_{/2}$ radians out of phase).

Circular polarization may be referred to as "right-hand" or "left-hand," depending on whether the helix describes the thread of a right-hand or left-hand screw, respectively

6.3   Importance of Chirality in biolgical Science

Why is the 3D-structure an important issue in drug design and in drug-receptor interaction?

Pharmacological activity of compounds(drugs) depend mainly on their interaction with biological matrices (drug targets), such as proteins (receptors, enzymes), nucleic acids (DNA and RNA) and biomembranes (phospholipids and glycolipids).

All these matrices have complex three-dimensional structures which are capable to recognize (bind) specifically the ligand (drug) molecule in only one of the many possible arrangements in the three dimensional space. It is the three-dimensional structure of the drug target that determines which of the potential drug candidate molecules is bound within its cavity and with what affinity. This control three dimensional shape of organic molecules (drugs) viewed from the perspective of their interaction with potential biological targets. Chirality in drugs most often arises from a carbon atom attached to 4 different group.

Understanding chirality is important for drug design. In order to target a particular biological receptor, the specificity and chirality of the recognition site must be correct. The fortuitous discovery of penicillin demonstrates the importance of chirality in drug design. The bacterial cell wall is composed of a polysaccharide component that is cross-linked by a short peptide. This peptide link contains D-alanine amino acids and is assembled by enzymatic reactions and not by the ribosome. The cross-linked cell wall macromolecule forms a stable shell that protects the bacteria from extreme conditions.

6.4   Importance of Chirality in agro-chemicals

Chirality in agrochemicals discuss the synthesis, activity and toxicology of chiral agrochemicals. The discovery that opposite enantiomers can have similarly diverse effects on plants and insects has had no less a revolutionary impact on the agrochemical industry. Insects use chiral chemical messengers (pheromones) as sex attractants and chemists have discovered that one of the enantiomers of the insect pheromone, olean, attracts male fruit flies, while its mirror image operates on the female of the species.

Pheromones play important roles in chemical communication among organisms. Various chiral and non-racemic pheromones have been identified since the late 1960's. Their enantioselective syntheses could establish the absolute configuration of the naturally occurring pheromones and clarified the relationships between absolute configuration and bioactivity. For example, neither the (R)- nor (S)-enantiomer of sulcatol, the aggregation pheromone of an ambrosia beetle Gnathotrichus sulcatus, is behaviorally active, while their mixture is bioactive. In the case of olean, the olive fruit fly pheromone, its (R)-isomer is active for the males, and the (S)-isomer is active for the females. About 140 chiral pheromones are reviewed with regard to their stereochemistry–bioactivity relationships. Problems encountered in studying chirality of pheromones were examined and analyzed to think about possible future directions in pheromone science.

Many Chirality in synthetic agrochemicals that are derived from natural products (e.g. pyrethroids) have chiral structures, the insecticidal activity being associated with one or more of the individual enantiomers. In the case of insecticidally active pyrethroids, this enantiomeric selectivity arises from the chiral nature of the site of action in the insect nervous system. Herbicidal 2-aryloxypropanoates exhibit their activity by inhibiting an acetyl-CoA carboxylase. The enantiomeric selectivity of these herbicides is thus understandable through the chiral nature of the enzyme molecule. Although some individual stereoisomers and enantiopure isomers of synthetic agrochemicals are now being marketed (e.g. pyrethroid insecticides, aryloxypropanoate herbicides and triazole fungicides), no single enantiopure isomer of chiral organo- phosphorus insecticides is commercially available.

6.5 Importance of Chirality in Flavours ,Food & Fragrances

The state of the art in research on chirality and odor is presented with special emphasis on synthetic and analytical methods. The syntheses of 1-octen-3-ol enantiomers, of chiral alkan-2-yl esters of "whisky lactone" and of chiral sulfur compounds known as characteristic components of the yellow passion fruit are reviewed. It is demonstrated that optically pure stereoisomers exhibit their own specific sensory characteristics. Therefore, stereoisomers of very high enantiomeric excess (ee) must be achieved for valid evaluations of their structure-function relationships. Analytical methods for stereo differentiation of chiral flavor compounds and recent developments in biotechnology are discussed.

Limonene is a colourless liquid hydrocarbon classified as a cyclic terpene. The more common D-isomer possesses a strong smell of oranges. It is used in chemical synthesis as a precursor to carvone and as a renewably based solvent in cleaning products.Limonene takes its name from the lemon as the rind of the lemon like other citrus fruits contains considerable amounts of this compound, which contributes to their odor. Limonene is a chiral molecule and biological sources produce one enantiomer: the principal industrial source, citrus fruit contains D-limonene   ((+)-limonene),   which   is   the   (R)-enantiomer . Racemic limonene is known as dipentene. D-Limonene is obtained commercially from citrus fruits through two primary methods: centrifugal separation or steam distillation.

Menthol

Menthol is a terpenoid alcohol with three chiral centres leading to eight possible stereoisomers (four enantiomeric pairs). Only the (-)-menthol enantiomer possesses the intense cooling and clean, desirable minty odour. For example, the (+)-menthol enantiomer is less cooling and possesses an undesirable musty off-note odour. This note is also present in racemic menthol.

The technique of enantioselective gas chromatography (GC) is described and applied for assigning absolute configuration of chiral natural compounds, which is strongly connected to differences in odor properties of their enantiomers. In addition, some recent results to facilitate the handling of GC–mass spectrometry data of known and unknown plant volatiles Enantioselective analysis of chiral constituents of essential oils and flavor and fragrance compounds has arrived at a point of perfection that is attributable to the work of a number of research groups in different disciplines (natural compound chemists, food chemists, perfumery chemists, and more recently, molecular biologists). However, the analysis of essential oils and flavor and fragrance compounds is not only a fascinating research field but also of considerable economical and social relevance. New regulations in the European Union on alleged allergenic essential oil constituents forces essential oil dealers and the perfumery industry to adopt more sophisticated analytical procedures including GC–MS and enantioselective GC. This will also be important in the coming decades, which will confront us with increasing numbers of genetically engineered plants with new fragrance and flavor characteristics.

The occurance of amino acids in a special optical configuration is also essential for sensory quality of food stuffs. Amino acids except for methionone depending on their configuration are neutral, sweet & bitter in taste. i.e. Four of them aspargine, tryptophane, tyrosine, isoleucine are characterized by bitter taste in their L-form & sweet taste in their D-form. The characteristic of numerous food products depends among other things on the content of the free amino acids including especially Glutamic acid. However only the L-form of the Glutamic acid is the carried of the taste defied as meat , brothy or umami taste. Both the D-form & racemates of the Glutamic acid do not have such a taste.

Protein plays a significant role in the process of taste reception. It is believed that the perception of bitter and sweet taste is dependent on the receptor connected to G-protein, gustducin,a chiral receptor such a protein may explain the enantioselective perception of taste sensations of for various configuration form of the amino acids.

However, academics did not generally accept the premise that optical enantiomers could have different odours until the early 1970'S when two research papers appeared that unequivocally proved that the carvone enantiomers were dramatically different in odour character. (-)-Carvone is the main constituent in spearmint oil and is the primary spearmint odour contributor. (+)-Carvone is the main constituent in caraway oil and is primarily responsible for its caraway odour.

6.6   Importance of Chirality in pharmaceutical Industry Enantiopure drug

An enantiopure drug is a pharmaceutical that is available in one specific enantiomeric form. Most biological molecules (proteins, sugars, etc.) are present in only one of many chiral forms, so different enantiomers of a chiral drug molecule bind differently (or not at all) to target receptors. Advances in industrial chemical processes have made it economical for pharmaceutical manufacturers to take drugs that were originally marketed as a racemic mixture and market the individual enantiomers either by specifically manufacturing the desired enantiomer or by resolving a racemic mixture. On a case-by-case basis, the U.S. Food and Drug Administration (FDA) has allowed single enantiomers of certain drugs to be marketed under a different name than the racemic mixture. Also case-by-case, the United States Patent Office has granted patents for single enantiomers of certain drugs. The regulatory review for marketing approval (safety and efficacy) and for patenting (proprietary rights) is independent and differs country by country.

Examples

The following table lists pharmaceuticals that have been available in both racemic and single-enantiomer form.

Racemic mixture	Single-enantiomer
Amphetamine (Benzedrine)	Dextroamphetamine (Dexedrine)
Bupivacaine (Marcain)	Levobupivacaine (Chirocaine)
Cetirizine (Zyrtec / Reactine)	Levocetirizine (Xyzal)
Citalopram (Celexa / Cipramil)	Escitalopram (Lexapro / Cipralex)
Ibuprofen (Advil / Motrin)	Dexibuprofen (Seractil)
Methylphenidate (Ritalin)	Dexmethylphenidate (Focalin)
Modafinil (Provigil)	Armodafinil (Nuvigil)

Ofloxacin (Floxin)	Levofloxacin (Levaquin)
Omeprazole (Prilosec)	Esomeprazole (Nexium)
Salbutamol (Ventolin)	Levalbuterol (Xopenex)
Zopiclone (Imovane)	Eszopiclone (Lunesta)

The following are cases where the individual enantiomers have markedly different effects:

- Ethambutol: Whereas one enantiomer is used to treat tuberculosis, the other causes blindness.
- Naproxen: One enantiomer is used to treat arthritis pain, but the other causes liver poisoning with no analgesic effect.
- Steroid receptor sites also show stereoisomer specificity.
- Penicillin's activity is stereodependent. The antibiotic must mimic the D-alanine chains that occur in the cell walls of bacteria in order to react with and subsequently inhibit bacterial transpeptidase enzyme.
- Only L-propranolol is a powerful adrenoceptor antagonist, whereas D-propranolol is not. However, both have local anesthetic effect.
- The L-isomer of Methorphan, levomethorphan is a potent opioid analgesic, while the D-isomer, dextromethorphan is a dissociative cough suppressant.
- (S)-(−) isomer of carvedilol, a drug that interacts with adrenoceptors, is 100 times more potent as beta receptor blocker than (R)-(+) isomer. However, both the isomers are approximately equipotent as alpha receptor blockers.
- The D-isomers of amphetamine and methamphetamine are strong CNS stimulants, while the L-isomers of both drugs lack appreciable CNS(central nervous system) stimulant effects, but instead stimulate the peripheral nervous system. For this reason, the Levo-isomer of methamphetamine is available as an OTC nasal inhaler in some countries, while the Dextro-isomer is banned from medical use in all

but a few countries in the world, and highly regulated in those countries who do allow it to be used medically.

- Ketamine is commonly composed of R & S enantiomers that have different dissociative and hallucinogenic properties, whereas the S enantiomer Esketamine is more potent in isolation as a dissociative.

Consequences and applications

Some of the applications of the Noble Laureates' pioneering work have already been discussed. It is especially important to emphasize the great significance their discoveries and improvements have for industry.   Industrial syntheses of new drugs are of major importance, but we may also mention the production of agro-chemicals including pheromones, flavours, fragrances and sweetening agents. Nobel Prize in Chemistry shows that the step from basic research to industrial application can sometimes be a short one. All around the world many research groups are busy developing other catalytic asymmetric syntheses that have been inspired by the Laureates' discoveries. Their developments have provided academic research with many important tools, thereby contributing to more rapid advances of research - not only in chemistry but also in materials science, biology and medicine. Their work gives access to new molecules needed to investigate hitherto unexplained and undiscovered phenomena in the molecular world.

REFERENCES

[1] J. Chem. Education, Vol.74, No.7, July 1997

[2] Lakhtakia, A. (ed.) (1990). *Selected Papers on Natural Optical Activity (SPIE Milestone Volume 15)*. SPIE.

[3] Pasteur, L. (1848). *Researches on the molecular asymmetry of natural organic products, published by Alembic Club Reprints (Vol. 14, pp. 1–46) in 1905*

[4] Pedro Cintas (2007), *Angewandte Chemie International Edition* **46** (22): 4016– 4024.

[5] Van't Hoff, Jacobus (September 1874). *Archives neerlandaises des sciences exactes et naturelles.* (9): 445-454.

[6] Fischer, Emil (1 October 1894). "Synthesen in der Zuckergruppe II". *Berichte der deutschen chemischen Gesellschaft* **27** (3): 3189–3232.

[7] Fischer, Emil; Hirschberger, Josef (1 January 1889). "Ueber Mannose. II". *Berichte der deutschen chemischen Gesellschaft* **22** (1): 365–376.

[8] Marckwald, W. (1904). "Ueber asymmetrische Synthese". *Berichte der deutschen chemischen Gesellschaft* **37**: 49.

[9] McKenzie, Alexander (1 January 1904). *Journal of the Chemical Society, Transactions* **85**: 1249.

[10] BIJVOET, J. M.; PEERDEMAN, A. F.; van BOMMEL, A. J. (1951). *Nature* **168**(4268): 271–272.

[11] Organic Chemistry (4th Edition) Paula Y. Bruice.

[12] Organic Chemistry (3rd Edition) Marye Anne Fox ,James K. White-sell.

[13] IUPAC, *Compendium of Chemical Terminology*, 2nd ed. (the "Gold Book") (1997). Online corrected version: (2006–) "Chirality".

[14] IUPAC, *Compendium of Chemical Terminology*, 2nd ed. (the "Gold Book") (1997). Online corrected version: (2006–) "Superposability".

[15] Gal, Joseph (2012). "The Discovery of Stereoselectivity at Biological Receptors: Arnaldo Piutti and the Taste of the Asparagine Enantiomers- History and Analysis on the 125th Anniversary". *Chirality* **24** (12)

[16] Theodore J. Leitereg, Dante G. Guadagni, Jean Harris, Thomas R. Mon, and Roy Teranishi (1971). "Chemical and sensory data supporting the difference between the odors of the enantiomeric carvones". *J. Agric. Food Chem.* **19** (4): 785.

[17] Lepola U, Wade A, Andersen HF (May 2004). *Int Clin Psychopharmacol* **19** (3): 149–55.

[18] Hyttel, J.; Bøgesø, K. P.; Perregaard, J.; Sánchez, C. (1992). *Journal of Neural Transmission***88** (2): 157–160.

[19] Jaffe, I. A.; Altman, K.; Merryman, P. (1964). *Journal of Clinical Investigation* **43** (10): 1869– 1873.

[20] Cronin, J. R. and S. Pizzarello. *Science* 275: 951-955.

[21] Pizzarello, A., Y. Huang, and M. R. Alexandre. 2008. meteorite. *Proc. Natl. Acad. Sci. U.S.A.* 105: 3700-3704.

[22] Herd, C. D. K., et al. 2011. *Science*332: 1304-1307.

[23] J. Martyn Bailey. 1998. *The FASEB Journal* 12: 503-507.

[24] Tamura, K. and P. R. Schimmel. 2006. *PNAS* 103: 13750-13752.

Tamura, K. 2008. *Nucleic Acids Symposium Series* 52: 415-416

[25] Hussain, T., S. P. Kruparani, B. Pal, A. Dock-Bregeon, S. Dwivedi, M. R. Shekar, K. Sureshbabu and R. Sankaranarayanan. 2006. *The EMBO Journal* 25: 4152-4162.

[26] Klussmann, M., H. Iwamura, S. P. Mathew, D. H. Wells Jr., U. Pandya, A. Armstrong, and D. G. Blackmond. 2006. *Nature* 441: 621-623.

[27] Lahav, N., D. White, and S. Chang. 1978. *Science* 201: 607.Orgel, L. E. 1989. *J. Mol. Evol.* 29: 465.Brack, A. 1993. *Pure Appl. Chem.* 65: 1103.

[28] Huber, C. and G. Wächtershäuser. 1998. *Science* 281: 670-672.

[29] Pizzarello, S. and A. L. Weber. 2004. *Science* 303: 1151.

[30] Engel, M. H. and S. A. Macko. 1997. *Nature* 389: 265-268.

[31] Mason, Stephen F. 1997. *Nature* 389: 804.

[32] Bailey, J. 1999. Polarized Stellar Light. *Science* 283: 1415. Bailey, J., A. Chrysostomou, J. H. Hough, T. M. Gledhill, A. McCall, S.

Clark, F. Menard, and M. Tamura. 1998. *Science* 281: 672-674.

[33] Kahr, B., and J.H. Freudenthal. 2008. *Chirality* 20: 973-7.

[34] Bauer, Eike B. (2012). *Chemical Society Reviews* **41** (8): 3153

[35]Comprehensive Asymmetric Catalysis (Jacobsen, Pfaltz, Yamamoto), Springer, 1999; b) Catalytic Asymmetric Synthesis, (Ojima), Wiley, 2000.

[36] M. Heitbaum, F. Glorius and I. Escher (2006). "Asymmetric Heterogeneous Catalysis".*Angewandte Chemie International Edition* **45** (29): 4732–4762.

[37] Asymmetric Catalysis on Industrial Scale, (Blaser, Schmidt), Wiley-VCH, 2004.

[38] Glorius, F.; Gnas, Y. (2006). *Synthesis* **12**: 1899–1930.

[39] Schmid, A.; Dordick, J. S.; Hauer, B.; Kiener, A.; Wubbolts, M. Witholt, B.*Nature* **409** (6817): 258–268.

[40] *Organocatalysis—after the gold rush* Søren Bertelsen and Karl Anker Jørgensen Chem. Soc. Rev., **2009**, 38, 2178–2189

[41] J.A. Osborn, F.H. Jardine, J.F. Young, and G. Wilkinson, J. Chem. Soc. A, 1711(1966)

[42] A. Miyashita, A. Yasuda, H. Takaya, K. Toriumi, T. Ito, T. Souchi and R. Noyori, J.Am.Chem. Soc. **102**, 7932 (1980).

[43] T. Ohta, H. Takaya and R. Noyori, Inorg. Chem., **27**, 566 (1988).

Abbreviations & Definitions

Optical activity : *Rotatory power of* plane of linearly polarized light as it travels through certain materials.

L- and D- : Levo rotatory and Dextro rotator.

S- and R- : Sinester and Rectus

R$_f$: Retention factor: Quantification of an amount that each component of a mixture travels.

Chromatography : *Chromatography* is the collective term for a set of laboratory techniques for separation of mixtures by virtue of differences in absorbency.

Spectroscopy : The Science deals the measurement of radiation intensity of light

TLC : Thin layer chromatography.

H LC : High performance Liquid Chromatography

NMR : Nuclear Magnetic Resonance.

IR : Infrared

EuFOD : Chemical compound used primarily as a shift reagent in NMR spectroscopy.

Homochirality : If all the constituent units are molecules of the same chiral form

Receptor : A molecule usually found on the surface of a cell, that receives chemical signals from outside the cell from external substances.

Stereochemistry: Involves the study of the relative spatial arrangement of atoms that form the structure of molecules and their manipulation in three dimension.

USC : *University of Southern California.*

ee's : Enantiomeric Excess.

SN^2 : Second order Nucleophilic substitution.

2D : Two dimension.

IU AC : International Union of Pure and Applied Chemistry.

OTC : *Over-the-counter.*

heromone : A *pheromone* is a chemical, a insect produces which changes the behavior of another insect of the same species.